AF599560

CREEPY BUT COOL

SNAKES

Julie K. Lundgren

A Crabtree Seedlings Book

CRABTREE
Publishing Company
www.crabtreebooks.com

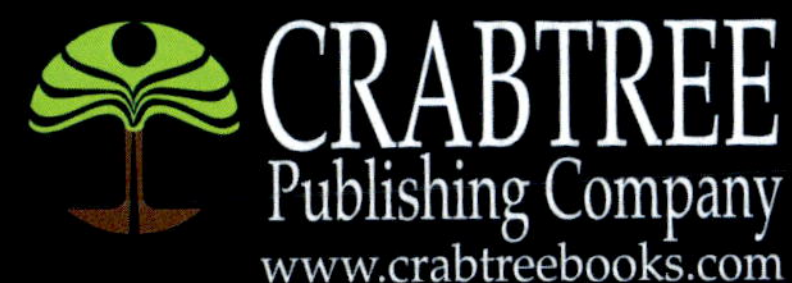

TABLE OF CONTENTS

SECRET SNAKES

More than 2,600 kinds of snakes live on Earth.

The largest snakes include the green anaconda of South America and the **reticulated** python. Most snakes live on land, but a few swim the seas.

Green anacondas are the heaviest snakes in the world.

Southern Pacific rattlesnake

CREEPY BUT COOL CLOSE-UP

Snake skin is dry and scaly, not slimy.

Like other reptiles, snakes have scales and use their surroundings to change their body temperature. They warm up in the Sun or cool down in the shade.

Snakes have different colors and patterns. Some use them for **camouflage**. Others wear bright stripes.

The poisonous coral snake's red, black, and yellow stripes act as a warning.

This green tree snake blends in among green leaves.

SLITHERING AND WRIGGLING

Snakes use their muscles and scales to wriggle in a **serpentine** motion, to the side, or like a caterpillar.

sidewinding adder

CREEPY OR COOL?

Why are snakes so flexible? They have a long spine and several hundred ribs.

FANGS OR SQUEEZES?

All snakes are **carnivores**. They may eat rodents, frogs, insects, birds, eggs, or other reptiles.

Snakes swallow their prey whole.

Some snakes kill with a quick bite containing **venom**.

Most snakes are harmless to people. Antivenin cures bites from venomous snakes.

Fangs inject venom.

Constrictors hug so tightly, the victim cannot breathe and its heart cannot beat.

Pythons coil and squeeze to kill their prey.

SUPER SNAKE SENSES

Snakes can smell and taste the air and track prey with their tongues. Their jawbones can feel vibrations from something walking nearby.

green rat snake

CREEPY OR COOL?

Pit vipers have special body parts that detect heat, allowing them to hunt in darkness.

The holes between the viper's eyes and mouth detect heat.

SNAKE LIFE CYCLE

Most snakes lay eggs. When they hatch, baby snakes look like tiny adult snakes.

Garter snakes are one kind of snake that do not lay eggs. They give birth to live young.

Growing snakes molt several times. They shed their too tight skin, revealing a new, larger skin.

Adults molt to replace old, worn skin.

Snakes may live ten to twenty years. These silent **predators** help people by eating insects, rats, and mice.

Red rat snakes help control rodents.

NAME THAT SNAKE!

Match each snake name below with the correct photo:

green rat snake

coral snake

garter snake

green anaconda

Southern Pacific rattlesnake

Answers: **a.** green anaconda **b.** green rat snake **c.** garter snake **d.** coral snake **e.** Southern Pacific rattlesnake

camouflage (KAM-uh-flahzh): Coloring or shape that allows an animal to blend in with its surroundings

carnivores (KAR-nuh-vorz): Animals that eat other animals

constrictors (kuhn-STRIK-terz): Snakes that kill by coiling around prey and squeezing it to death

predators (PREH-duh-terz): Animals that hunt other animals

reticulated (ruh-TIK-yu-lay-tuhd): Having a pattern of markings that looks like a net

serpentine (SER-puhn-teen): Curved in the shape or pattern of the letter "S"

venom (VEN-um): Poison used to stun or kill prey, delivered by a bite

INDEX

School-to-Home Support for Caregivers and Teachers

This book helps children grow by letting them practice reading. Here are a few guiding questions to help the reader build his or her comprehension skills. Possible answers appear here in red.

Before Reading

- **What do I think this book is about?** *I think this book is about creepy snakes. I think this book is about different kinds of snakes.*
- **What do I want to learn about this topic?** *I want to learn how to avoid snakes. I want to learn how to identify poisonous snakes.*

During Reading

- **I wonder why...** *I wonder why there are so many different kinds of snakes on Earth. I wonder why some snakes can swim.*
- **What have I learned so far?** *I have learned that snakeskin is dry and scaly, not slimy. I have learned that snakes can smell and taste the air, and track prey with their tongues.*

After Reading

- **What details did I learn about this topic?** *I have learned that there are more than 2,600 kinds of snakes living on Earth. I have learned that all snakes are carnivores that catch and eat other animals.*
- **Read the book again and look for the glossary words.** *I see the word* ***camouflage*** *on page 8, and the word* ***venom*** *on page 14. The other glossary words are found on page 23.*

Library and Archives Canada Cataloguing in Publication

Title: Snakes / Julie K. Lundgren.
Names: Lundgren, Julie K., author.
Description: Series statement: Creepy but cool | "A Crabtree seedlings book". | Includes index. | Previously published in electronic format by Blue Door Publishing FL in 2015.
Identifiers: Canadiana (print) 20210201908 | Canadiana (ebook) 20210201916 | ISBN 9781427161697 (hardcover) | ISBN 9781427161819 (softcover) | ISBN 9781427161932 (HTML) | ISBN 9781427162052 (EPUB) | ISBN 9781427162175 (read-along ebook)
Subjects: LCSH: Snakes—Juvenile literature.
Classification: LCC QL666.O6 L86 2022 | DDC j597.96—dc23

Library of Congress Cataloging-in-Publication Data

Names: Lundgren, Julie K., author.
Title: Snakes / Julie K. Lundgren.
Description: New York : Crabtree Publishing, [2022] | Series: Creepy but cool - a Crabtree seedlings book | Includes index.
Identifiers: LCCN 2021018425 (print) | LCCN 2021018426 (ebook) | ISBN 9781427161697 (hardcover) | ISBN 9781427161819 (paperback) | ISBN 9781427161932 (ebook) | ISBN 9781427162052 (epub) | ISBN 9781427162175
Subjects: LCSH: Snakes--Juvenile literature.
Classification: LCC QL666.O6 L83 2022 (print) | LCC QL666.O6 (ebook) | DDC 597.96--dc23
LC record available at https://lccn.loc.gov/2021018425
LC ebook record available at https://lccn.loc.gov/2021018426

Crabtree Publishing Company
www.crabtreebooks.com 1–800–387–7650
Print book version produced jointly with Blue Door Education in 2022

Written by Julie K. Lundgren
Print coordinator: Katherine Berti
Printed in Canada/042023/CPC20230406

Photo credits: www.shutterstock.com - www.istock.com. Cover © Audrey Snider-Bell. page 2-3: istock.com/Aulia Maghfiroh. page 4-5 anaconda © Patrick K. Campbell. page 6-7 © Rusty Dodson, close-up snake skin © Audrey Snider-Bell. page 8-9 coral snake © Patrick K. Campbell, tree snake © Zdenek Rosenthaler. page 10-11 © Arno Dietz. page 11 inset © srdjan draskovic,. page 12-13 © Trahcus. page 18 © Maria Dryfhout. page 15 © Andrew Burgess. page 16-17 © reptiles4all . page 17 inset © H. Krisp wikimedia commons. page 18-19 © Heiko Kiera, page 29 gartner snake © Matt Jeppson page 20 © manit321. page 21 © Patrick K. Campbell. www.dreamstime.com

Published in the United States
Crabtree Publishing
347 Fifth Ave.
Suite 1402-145
New York, NY 10016

Published in Canada
Crabtree Publishing
616 Welland Ave.
St. Catharines, Ontario
L2M 5V6